TIME UNRAVELED: A SIMPLE GUIDE TO QUANTUM TRAVEL

Exploring the Boundaries of
Time, Space, and Reality

Alireza Minagar

Amazon

Copyright © 2024 Alireza Minagar

All rights reserved

The characters and events portrayed in this book are fictitious. Any similarity to real persons, living or dead, is coincidental and not intended by the author.

No part of this book may be reproduced, or stored in a retrieval system, or transmitted in any form or by any means, electronic, mechanical, photocopying, recording, or otherwise, without express written permission of the publisher.

ISBN-13: 9798343885545

Cover design by: Art Painter
Library of Congress Control Number: 2018675309
Printed in the United States of America

CONTENTS

Title Page
Copyright
Preface
Prologue   1
Chapter 1: Introduction to Time Travel and its Mysteries   3
Chapter 2: Understanding Time   7
Chapter 3: The Foundations of Quantum Mechanics   10
Chapter 4: Einstein's Theory of Relativity and Time Dilation   15
Chapter 5: The Connection Between Quantum Mechanics and Time Travel   22
Chapter 6: Theoretical Models of Time Travel   26
Chapter 7: Practical Considerations and Challenges   29
Chapter 8: Philosophical Implications of Time Travel   31
Chapter 9: Future of Time Travel Research   33
Chapter 10: The Multiverse Theory: Parallel Universes and Time Travel   39
Chapter 11: The Grandfather Paradox and Other Time Travel Dilemmas   43
Chapter 12: Eleven Dimensions of Reality   46
Chapter 14: Ethical Implications of Time Travel: Should We Do It?   52

Chapter 15: Time Travel and Sleep: The Dream of Timeless   55
Journeys

Chapter 16: Conclusion   59

PREFACE

Time travel has long captivated human imagination, weaving its way through the realms of science fiction, philosophy, and even serious scientific discourse. The concept of moving backward or forward in time challenges our most fundamental understandings of reality. Yet, as we have advanced in physics, what once seemed like pure fantasy now lingers at the fringes of possibility.

Isaac Newton, one of history's most brilliant minds, viewed time as an absolute, unchanging entity-a steady river flowing in one direction. According to Newtonian mechanics, time was independent of space, fixed like an invisible framework that governed all motion in the universe. For centuries, this view shaped our understanding of reality.

Then came Albert Einstein, whose theory of general relativity dramatically altered the landscape of physics. He merged space and time into a unified concept known as spacetime, a flexible fabric that could be warped by mass and energy. With this revolutionary idea, the door opened to new possibilities for how time could behavepossibilities that include phenomena like wormholes and black holes. Wormholes, theoretical bridges connecting distant points in spacetime, offer a tantalizing prospect: the potential to travel not just across vast stretches of space, but also through time itself.

But Einstein's contributions did not stop there. General relativity also introduced the concept of closed timelike curves, which in theory could enable a return to the past. Such ideas challenge the very nature of causality and raise famous paradoxes, like the grandfather paradox, where a time traveler might prevent their own existence. Despite these philosophical conundrums, physicists continue to refine mathematical models to explore whether time travel could indeed be possible.

Quantum mechanics adds further complexity. Unlike the classical physics of Newton and Einstein, quantum physics reveals a world that is probabilistic rather than deterministic. Particles can exist in multiple states at once through superposition, and the phenomenon of entanglement suggests that particles, even when separated by great distances, can remain interconnected across space and time. Some researchers propose that these quantum properties could lead to alternate timelines or parallel universes, offering yet another speculative avenue for time travel.

Though many questions remain unanswered, these groundbreaking theories offer glimpses of a future where time travel may not be limited to the pages of science fiction. With new advances in quantum mechanics and our ever-evolving understanding of spacetime, we may one day unlock the secrets of time itself.

Disclaimer

This book is intended for informational and educational purposes only and is meant to be a short, fun read for relaxation, not a comprehensive textbook on physics or the mechanics of time travel. The theories discussed are speculative and remain within the domain of theoretical physics. The author and publisher have made their best efforts to ensure accuracy and clarity; however, they do not accept any responsibility for any errors, interpretations, actions, or consequences arising from the content presented in this book.

All rights reserved. No part of this book may be reproduced, distributed, or transmitted in any form or by any means without the prior written permission of the publisher, except for brief quotations in critical reviews or other non-commercial uses permitted by copyright law.

PROLOGUE

Time has always been one of humanity's most enduring mysteries. From the steady tick of a clock to the vast spans of ages, past and future, time shapes our existence in ways we barely understand. Yet, as with many things in the universe, the more we study time, the more it seems to unravel before revealing unexpected complexities and possibilities far beyond our everyday experience. With the rise of quantum physics and relativity, time is no longer a mere backdrop to our lives; it becomes a force we may one day learn to navigate.

In *Time Unraveled: A Simple Guide to Quantum Travel*, we journey through some of the most exciting ideas that modern science has to offer. Concepts once confined to the realm of science fiction—like traveling to the past, jumping across alternate timelines, or even existing in multiple states at once—are becoming legitimate topics of scientific inquiry. With the development of theories like quantum entanglement, time dilation, and the Many-Worlds Interpretation, scientists are beginning to glimpse how time may be more malleable than we once believed.

This book is not just for the physicist or the theorist. It is for anyone interested in time and how we might one day transcend its boundaries. The pages ahead will explore how wormholes might connect distant points in spacetime, how quantum particles defy the classical notion of time's arrow, and how the multiverse may offer us a roadmap to understanding alternate versions of our reality.

We will discuss the paradoxes, possibilities, and ethical implications that arise when we entertain the idea of quantum travel. Along the way, we will address key themes that underpin this fascinating field—uncertainty and probability, the bending of spacetime, and the relationship between time and identity.

As we unravel the complexities of quantum mechanics and the mysteries of time itself, remember that while the science may be complicated, the desire to understand and explore is simple—and deeply human. Let this guiding light be your leading stride into the remarkable sphere of quantum travel. Together, we will unravel the threads of time itself and open our minds to the staggering possibilities that await.

CHAPTER 1: INTRODUCTION TO TIME TRAVEL AND ITS MYSTERIES

The Model of Time Travel

Time travel has stunned human creativity for centuries, inspiring limitless stories, movies, and scientific discussions. The possibility of traversing through time, whether to witness past historical events or glimpse the future, continues to spark curiosity. Whereas time travel has long been a core of science fiction, certain principles within physics suggest that it may be more than just fantasy. Specifically, the fields of quantum mechanics and Einstein's theory of relativity offer insights that bring time travel closer to scientific consideration.

At its core, the theory of time travel revolves around the makeup of time itself. Traditionally, time has been viewed and researched as a progressive and linear flow of events from the past through the present and into the future. However, Einstein's theory of relativity challenges this notion, showing that time is not absolute but can be manipulated by certain other dynamics such as speed and gravity. This phenomenon, known as time dilation, implies that time moves differently depending on how fast an object is traveling or how close it is to a massive object. For example, astronauts traveling near the speed of light would experience time at a slower rate than those on Earth. This shows that travel into the future, at least theoretically, is possible.

Travelling back in time, on the other hand, poses greater arguments and obstacles. While traveling into the future is rooted in established physics, moving backward through time introduces logical paradoxes. The famous "grandfather paradox" is a well-known example: if a time traveler returns in time and blocks their grandfather from meeting their grandmother, the traveler would never be born, creating a contradiction. Some theories effort to solve these puzzles by suggesting the existence of alternate timelines or branching universes. In these models, changes to the past create new, divergent realities without altering the original timeline.

Quantum mechanics adds complexity to this discussion. At the quantum level, particles act in ways that disregard classical physics. The many-worlds analysis of quantum mechanics suggests that every decision or event may spawn a new universe, allowing for multiple realities to exist simultaneously. This theory could provide a framework for time travel without paradoxes, as a traveler could enter an alternate timeline rather than affecting their own.

Though we are far from developing time travel technology, the principles of quantum mechanics and relativity encourage us to think beyond conventional boundaries. By continuing to explore these scientific concepts, we may one day understand the mechanisms of time more deeply and even contemplate traveling through its enigmatic dimensions.

Historical Perspectives on Time Travel

The appeal of travelling through time, to the past or future, is not a contemporary experience. Throughout history, ancient cultures viewed time in ways that transcend the linear progression we commonly accept today. For example, many believed in cyclical time—an endless loop of life, death, and rebirth—reflected in their myths and religious practices. These early concepts laid the groundwork for more sophisticated explorations of time by philosophers during the Enlightenment. Immanuel Kant, for instance, saw time as a construct of human perception, shaping the way we experience the world rather than being an objective reality.

By the 19th century, significant advances in physics began to reshape our understanding of time. Thermodynamics introduced the concept of entropy, emphasizing the unidirectional flow of time, often referred to as the "arrow of time." At the same time, literature started to reflect these evolving ideas. H.G. Wells' classic novel *The Time Machine* (1895) delved into the moral and scholarly consequences of time travel, sparking public interest in the subject.

The 20th century marked a turning point with Einstein's special theory of relativity, which exhibited that time is relative, not constant, and absolute. This breakthrough altered our understanding of time and space as interwoven dimensions. The era also saw the rise of popular science literature, bringing these complex ideas to a broader audience.

As quantum mechanics emerged, physicists began to propose even more speculative ideas, such as wormholes and closed timelike curves, potential pathways for traversing time. These concepts, though mathematically intriguing, raise fundamental questions about causality and the outcomes of modifying the past. Despite these challenges, public fascination with time travel continues to grow, supported by scientific inquiry and popular culture alike.

Popular Culture and Time Travel

Time travel has become a staple of popular culture, providing a canvas for exploring complex scientific themes and philosophical questions. Movies, television shows, and books depict time travel as both an exciting adventure and a source of profound dilemmas, often drawing on scientific principles to lend credibility to their stories. Works like *Back to the Future*, *Doctor Who*, and *The Time Machine* not only entertain but also introduce audiences to fundamental ideas rooted in quantum mechanics and relativity.

Popular culture frequently incorporates the concept of time dilation from Einstein's theory of relativity, often exaggerating it for dramatic effect. In these stories, characters may experience time differently from those around them, with the scientific principle serving as a backdrop for narratives that challenge the characters' perceptions of reality.

The grandfather paradox is another common theme in time travel stories, where characters grapple with the consequences of altering past events. These narratives engage audiences in philosophical questions about causality and the potential ripple effects of changing history.

Moreover, popular culture often draws from scientific

advancements, incorporating ideas like wormholes and quantum particles to create imaginative methods for traversing time. This blending of science and fiction not only entertains but also sparks curiosity about the underlying principles, inspiring audiences to explore the mysteries of time and space further.

Time Travel in Literature and Imagination

Time travel has long been a compelling theme in literature, offering authors a tool to explore deep philosophical questions about fate, free will, and the consequences of human actions. From H.G. Wells's *The Time Machine* to Audrey Niffenegger's *The Time Traveler's Wife*, these stories challenge readers to imagine what it would be like to alter the past or peek into the future. Time travel fiction often explores the tension between our desire to control events and the unpredictable nature of outcomes, using time as a lens through which we reflect on morality and existence. While rooted in imagination, time travel literature frequently draws from scientific theories, particularly those related to Einstein's theory of relativity. Authors use the idea that time is a flexible dimension, shaped by speed and gravity, to ground their stories in real physics, even as they stretch the boundaries of possibility. In recent years, quantum mechanics has added a new layer to time travel fiction, with the notion of multiple realities and alternate timelines creating more intricate and speculative narratives.

These imaginative landscapes serve as reflections of societal anxieties and aspirations, with dystopian or utopian visions of the future offering commentary on the human condition. By using time travel as a narrative device, authors invite readers to explore the implications of their choices and the impact of technology on society, all within the framework of speculative fiction.

CHAPTER 2: UNDERSTANDING TIME

What is Time?

Along with things such as the body and the environment, in one way or another time is surely one of the most 'structural' parts of that elephant called 'us', something bordering on nature itself. Yet it is hard to point to something so elusive, as if it did not really exist at all or certainly is very difficult to describe. Traditionally, since Isaac Newton's time, we think of time as a succession of past-through-present-into-future. Modern physics paints a much subtler picture. Albert Einstein's theory of relativity showed that what we commonly assume about time – that it is an abstract, inter-Coordinatable universal quantity in which things are sequenced – is not actually the case. Time, according to relativity, is relative rather than absolute. From an observer's point of view, it can be extended or contracted according to her speed or position relative to another massive body, something known as time dilation. Time dilation has been observed in many experiments, most famously in studies that involve putting atomic clocks at different altitudes or distances from the Earth, measuring tiny variations in the rate at which they tick as related to each other.

After relativity came quantum mechanics. At the quantum level, particles do not follow the rules of classical physics, and time, especially, loses its arrow. Superposition, for one, suggests particles can occupy multiple states at any given time, which means they might exist outside and across linear time.

Philosophically speaking, the question of whether or not time is real has been at the heart of debates from Aristotle to the present age in different ways. In the popular view, explicitly stated by the Buddha, only the moment you are experiencing is real. But the realist's view that all times – past, present, and future – are real claims an admission among philosophers; it is what is sometimes called eternalism. What happens in science is that these philosophical debates are increasingly coming into contact and even overlap with scientific accounts of reality. The theories of time that we construct through physics tend to include a separation of times from the space that we experience, adding in new levels of time altogether.

The Nature of Time in Physics

Physicists often refer to time as the fourth dimension, alongside the three dimensions of space. This concept was fundamentally reshaped by Einstein's theory of special relativity, which demonstrated that time is not just a sequence we experience, but it is inextricably linked with the three spatial dimensions to form a four-dimensional continuum known as spacetime. This breakthrough revealed two surprising truths about time that defy our everyday understanding. First, time is not an absolute concept that arises when we fail to account for how we and others move relative to one another. Secondly, time is not uniform. It does not tick at the same rate everywhere, as typical clocks run faster or slower depending on the strength of the gravitational field where they are located.

Einstein's general theory of relativity introduced the idea that massive objects, even inert ones like planets or stars, warp the fabric of spacetime itself. This curvature affects the passage of time, particularly in regions of intense gravitational strength. For example, time passes more slowly in the more powerful gravitational domains, a phenomenon known as gravitational time dilation. This prediction has been rigorously confirmed by experiments, such as placing atomic clocks at different altitudes: clocks closer to Earth's surface (where gravity is stronger) tick more slowly compared to those further away.

At the quantum level, even stranger things happen to time. Based on the quantum mechanics, particles can be existent in a superposition of states, meaning they occupy multiple states or positions until observed. This challenges our traditional, linear concept of time, suggesting that the arrow of time may not always follow the before-and-after pattern we are used to. In fact, quantum events are not always neatly ordered in a sequential timeline. Scientists are investigating whether time might be an emergent property, something that arises from more fundamental, quantum processes. If this is true, it may indicate that spacetime and causality are constructed from even more elementary principles that we are only beginning to understand.

By exploring relativity and quantum mechanics, physicists are beginning to stretch and reshape our understanding of time itself. While time travel remains within the realm of speculation and theoretical models, discoveries in these fields offer tantalizing glimpses of how time might be manipulated in ways once thought impossible.

Still, by pursuing the challenges of relativity and quantum mechanics, physicists may be starting to bend time itself. If there is not already a time-travel drawing board in a private room at the CERN headquarters in Geneva, there should be. And thanks to recent developments in physics, the plans already provide tantalizing blueprints for just how time might be taken for a spin.

CHAPTER 3: THE FOUNDATIONS OF QUANTUM MECHANICS

Basic Principles of Quantum Mechanics

Quantum mechanics is one of the most creative areas of modern physics. Concerning reality at the smallest scales, it says that things are probabilistic instead of certain. Classical mechanics characterizes the laws of motion of objects that we can see using our unaided senses, such as balls and hoops, where a force acting on a body gives it a definite response. As a deterministic theory, classical mechanics predicts exactly what nature must do when, if it is to be consistent. That theory's idea of time is progressive, meaning that it marches forward from earlier to later. In classical mechanics, the future is determined by the past, and familiar to us since we can describe it using our own experience. In this field, time appears to flow and is singular, passing from the present to the future, while reminding us of those science-fiction depictions where time reverses. Interestingly, classical mechanics itself is time-reversible at a fundamental level, even though we do not experience time as such.

Another key aspect of quantum mechanics, wave-particle duality, indicates that particles such as electrons can show wave-like properties or particle-like properties, depending upon how they are measured. This inherent duality between particles and waves requires that we reassess the deeply held concepts of traditional physics, some of which have implicated matter and energy behaving in regularly predictable ways.

The **uncertainty principle**, devised by Werner Heisenberg in 1927, is intricately linked to quantum mechanics. It asserts that it is impossible to precisely determine both the position and momentum of a particle at once. The more information we have about one property, the less precisely we can know the other. This inherent uncertainty is not a result of measurement limitations but is instead a quantum feature that invokes probabilism into the very fabric of reality.

To make the disparity between classical and quantum mechanics even clearer, the principle of **superposition** means that a quantum particle, such as an electron, can have several positions—or any other physical quality—simultaneously until it is observed, after which it selects one of them in a phenomenon called **wave function collapse**. It is the superposition that permits quantum computing, which means that a qubit can, simultaneously, be both a 0 and a 1. This enables computations of a complexity that are unheard of in the classical realm.

Also associated with superposition is an effect known as **quantum entanglement**. When two particles become entangled, their states become interconnected in such a way that the properties of one particle instantly imprint the properties of the other, no matter how distant they are separated by space. Einstein referred to this as "spooky action at a distance." This correlational effect, a form of Einsteinian spookiness, involves the violation of classical notions of locality and causality. When two quantum particles become entangled, it seems that something other than just a mechanical propagation of forces is occurring between them at the quantum scale.

More recent ideas about **quantum emergence** align with speculations on how quantum mechanics might apply within a theory of quantum gravity, the goal being to unify it with Einstein's general theory of relativity. Some physicists are examining whether a quantum description of time might be an emergent property of other quantum processes—an altogether different picture of time than that of the classical block universe; one that retains discrete jumps but without the necessity of branching histories. These move us to the outer limits of current physical theory.

Key Experiments and Discoveries

Several key experiments underpin the theoretical framework of quantum mechanics and have reshaped our understanding of the universe. One such experiment is the **Michelson-**

Morley experiment (1887), originally intended to detect the "luminiferous ether," a substance contemplated to transfer light waves. Instead, it demonstrated that the speed of light is constant in all directions, laying the groundwork for Einstein's **special theory of relativity** and revolutionizing the understanding of spacetime.

Einstein's **special relativity**, proposed in 1905, introduced the concept of **time dilation**, where time passes at dissimilar rates for observers moving at dissimilar velocities. This has been experimentally verified through studies, such as those involving atomic clocks situated on airplanes. These clocks, when in motion, tick slightly slower than stationary clocks, confirming the malleability of time—a key insight into the flexible nature of spacetime and time's non-absolute behavior.

Another pivotal breakthrough came in 1935, when Einstein and Nathan Rosen proposed the concept of **wormholes**—hypothetical tunnels that could connect distant points in spacetime, potentially allowing for time travel. Though still speculative, wormholes offer a tantalizing theoretical pathway for traversing time, sparking scientific and philosophical debates about the nature of causality and the possibility of altering the past.

In the 1950s, Hugh Everett III introduced the **many-worlds interpretation** of quantum mechanics, proposing a multiverse in which every quantum event advances to the formation of parallel realities. In the context of time travel, this theory implies that changes to the past might design new, parallel universes rather than modifying the original timeline. This interpretation of quantum mechanics has inspired not only scientific inquiry but also numerous works of fiction exploring the multiverse and alternate realities.

Quantum mechanics also plays a vigorous role in modern technological advancements, specifically in the growth of **quantum computing**. Unlike classical computers, which process data in binary form (as 0s and 1s), quantum computers utilize **qubits**, which can exist in a superposition of both 0 and 1 at the same time. This gives the power to the quantum computers to process information exponentially quicker than conventional computers, opening new possibilities in fields like cryptography, artificial intelligence, and quantum simulations. The future of quantum mechanics, especially in its application to innovative technologies, promises to reshape our scientific and technological landscape.

Quantum Theory vs. Classical Physics

Quantum theory and classical physics represent two distinct ways of understanding the universe. Classical physics, rooted in deterministic laws, works exceptionally well for the macroscopic world—planets orbiting the sun, the motion of a pendulum, or an apple falling to the ground. If the preliminary specifications of a system are recognized, classical physics can predict the imminent behavior of that system with accurateness.

Quantum mechanics, however, shattered this deterministic worldview. At the atomic and subatomic levels, particles do not behave in accordance with classical expectations. Instead, they occur in states of probability, as described by the **wave function**, until they are detected. This probabilistic nature contrasts sharply with classical physics, where objects always have well-defined properties.

In classical mechanics, time is regarded as a constant steady, unchanging framework against which events unfold. **Newton's laws of motion** assumed that time flows uniformly and independently of objects and forces. Quantum mechanics, in contrast, presents a far more complex relationship between time and matter. In the quantum realm, particles can become entwined, meaning their properties remain linked regardless of the distance parting them. This phenomenon, confirmed by numerous experiments, suggests a **non-local connection** between particles that defies classical concepts of space and time.

Quantum mechanics also opens new doors to understanding time travel. **Quantum tunneling**, where particles pass through barriers instantaneously, hints at the possibility that particles— and even larger objects—might "leap" through time. This provides a fascinating, albeit theoretical, glimpse into the potential of quantum mechanics to challenge our orthodox interpretation of time and space.

The pursuit to merge **quantum mechanics** with **general relativity** remains one of the biggest obstacles and concepts in modern physics. Researchers are working to bridge the gap between the macroscopic laws governing planets and galaxies and the quantum rules governing subatomic particles. If successful, this unified theory—often referred to as **quantum gravity**—could revolutionize our understanding of time, space, and the very nature of reality.

CHAPTER 4: EINSTEIN'S THEORY OF RELATIVITY AND TIME DILATION

Special Relativity Explained

Albert Einstein's theory of special relativity, proposed in 1905, is one of the most fundamental insights into the nature of reality. It does away with the classical idea of final and absolute, independent space and time that was put forth by Sir Isaac Newton in the Principia of 1687, and replaces it with spacetime, wherein space and time are not separate and independent, but are both aspects of a unified manifold. Time, in contrast to the Newtonian conception, is no longer absolute: the same event acquires various times depending on the velocity and position of an observer in space. This perspective dramatically transformed physics and had far-reaching consequences for many other fields of research.

The most famous consequence of special relativity – time dilation – predicts that the closer you get to the speed of light relative to someone standing still (or travelling more slowly), the slower your clock will tick in that person's reference frame. This result, though unintuitive, has been observed in the laboratory consistently repeatedly. The classic demonstration of this notion is the so-called twin paradox: take two identical twins. One stays on Earth, while the other climbs into a space probe and travels near the speed of light for several decades. The return of the space traveler shows that he arrives at Earth with a younger appearance than his Earth-bound brother. Motion causes time, and this means that all real-world motion talks about time science-fictional space flight is only one setting in which time dilation is important. In the context of the definition of everyday life, motion is so slow that no such effect can be observed. But when the speeds increase towards the speed of light, the impact is substantial. Indeed, in some of the most important sub-fields of physics, such as particle and nuclear physics, or astrophysics, the effect is undeniably real. It is key to understanding the fundamental properties of matter (or energy) in the form of fast-moving particles. We have already encountered the notion of time moving more slowly for objects that travel close to the speed of light. This is precisely what happens to particles produced at higher-energy accelerators, such as the Large Hadron Collider at CERN near Geneva. Since these particles have very brief lives, their ability to decay depends upon how much time they have while flying about inside the particle collider, through which they pass. If time moves more slowly for these particles than for an observer in the laboratory, it follows that particles close to the speed of light also have much longer lives than if they had been at rest – a difference that can be measured in the laboratory.

Another cornerstone of special relativity is the celebrated equation $E=mc^2$, which expresses the equivalence of mass and energy. This equation indicates that mass can be changed into energy, and the other way around, with the speed of light squared acting as the conversion factor. This insight laid the foundation for nuclear energy and has helped scientists understand processes as diverse as nuclear reactions in stars and the creation of particle-antiparticle pairs in high-energy environments. In fact, every time a star like our Sun generates energy, it is converting a tiny amount of mass into an enormous amount of energy, following Einstein's principle.

The speed of light, approximately 299,792 kilometers per second (or 186,282 miles per second), serves as the extreme speed limit in the universe (or at least this is our present impression). Nothing with mass can travel at or above this pace because, as objects approach it, their mass increases, requiring an infinite amount of energy to keep accelerating. This speed limit has profound consequences for the nature of casualty and how we understand the transmission of information through space and time. In a sense, the speed of light represents the boundary beyond which the normal rules of space and time begin to break down, ushering us into the realms of quantum mechanics and general relativity.

As profound as special relativity is, it is important to note that it applies primarily to objects moving in uniform motion—those not under the influence of forces such as gravity. To address situations involving gravity, Einstein developed the theory of general relativity, which expands upon special relativity to incorporate the effects of gravitational forces on the structure of spacetime.

General Relativity: Gravity and Spacetime

In 1915, Einstein published his theory of general relativity, which primarily altered our understanding of gravity. While Isaac Newton had previously described gravity as a force between two masses, Einstein suggested that gravity is the result of the curving of spacetime. Immense objects, such as stars and planets, distort the spacetime around them, creating a kind of "dip" in the fabric of spacetime. Other objects, like moons or satellites, follow these curves, which is why they are pulled towards the massive object. This distortion of spacetime is what we remark as the push of gravity.

General relativity posits that spacetime is a four-dimensional continuum, in which time is added to the three spatial dimensions of length, width and height. This fabric is warped (curved) by the presence of mass. The more mass, the more curvature there is to the spacetime in question. Essentially, gravity is a manifestation of the mass-generated curvature of spacetime, and not the instantaneous action-at-a-distance described by Newton. This model predicted numerous phenomena that had been 'fudged' by Newtonian mechanics, such as the precise orbital motions of the planets, as well as the way light is bent by massive objects. Light passing one side of a massive object appears further away on the other side by an amount that depends on the amount of mass causing the deflection. This elegant gravitational lensing phenomenon has been confirmed from both earth-based telescopes and space observatories.

Under the influence of curved spacetime, light takes a bent path. The presence of light rays bending around massive intervening bodies, such as other galaxies or black holes, is why we call this phenomenon gravitational lensing. The result is that from our perspective here on Earth, a distant light source – a dim star or galaxy, for instance – appears warped or magnified. Gravitational lensing has now been confirmed in countless examples. It functions as a tool for astronomers to probe newly detectable phenomena, such as the presence of dark matter, black holes and the evolution of the large-scale structure of the cosmos. Gravitational lensing can allow us to map out the distribution of dark matter, which does not emit light. It produces no intrinsic signal of its own, but we know that it alone, in the correct amounts, can deform spacetime.

One of the most dramatic predictions of general relativity is the existence of black holes, regions of spacetime in which gravity is so strong that not even light can escape. Supermassive black holes are thought to form when giant stars collapse from their own gravity, crushing all their matter into a single point, called a singularity. The margin of a black hole, outside which nothing may escape, is called the event horizon. The curvature of spacetime is so extreme inside the event horizon that all paths are bent inward toward the singularity, including those taken by light.

Black holes evaluate our understanding of the universe by challenging our method for combining the two best theories we have for describing nature. They unite general relativity and quantum mechanics in a way that reveals a hole can neither fully describe. Would a black hole fundamentally change how people perceive the universe? Indeed, the grand vision of the late Stephen Hawking has been that black holes might be the key to unifying the two fundamental theories we have for describing the forces of nature – Albert Einstein's general relativity for gravity, and quantum mechanics for trivial things.

The real world implications of general relativity can be seen in the functioning of the Global Positioning System (GPS), which depends on satellites orbiting around the Earth. Satellites orbiting high in the sky experience time difference from objects on the Earth's surface both because of their speed and their distance from the gravitational field of Earth. GPS systems would lose all accuracy within hours if they did not compensate for this time dilation, according to general relativity.

Implications of Relativity on Time Travel

Einstein's relativity theories not only describe the behavior of space, time, and gravity but also open the door to intriguing possibilities, such as time travel. Time dilation—the slowing of time due to high velocity or strong gravitational fields—suggests that traveling to the future is theoretically possible. Astronauts traveling near the speed of light, or those near a massive object like a black hole, could experience time differently from those on Earth.

Next, the alternative idea: black holes are associated with a less famous, more speculative aspect of relativity in their own right – so-called Einstein-Rosen bridges, or wormholes. A wormhole is a theoretical 'tunnel' in spacetime that could connect two distant points in space or time. The mathematics of general relativity admits the concept, at least in principle – but it generally requires 'exotic matter', which can have a negative energy density, to hold a wormhole open. So far there is no laboratory evidence for exotic matter, nor for stable wormholes, but the idea continues to fascinate and inspire a plethora of science fiction stories.

But if one were to travel backwards – that is, to backward time travel – then paradoxes of a different type arise, such as the 'grandfather paradox' suggesting that a traveler going back in time could kill her grandfather before he can father her mother, and thus prevent her own existence. In the case of backward time travel, physicists have proposed several ways to avoid or resolve paradoxes based on different scenarios. One is the just-mentioned idea that changes in the past create parallel realities, or parallel universes. Another is the concept of many-worlds where there are many versions of reality, seen as simultaneous and shareable. But while backward time travel is only speculation, forward time travel through time dilation is quite real. Time dilatation was predicted by Einstein's relativity, and certain forms of it have been measured in particle accelerators and in experiments involving atomic clocks.

This interplay between time, space, and gravity continues to be a spur to scientific investigation and philosophical discussion. Einstein's relativity remains a framework for exploration, as physicists begin to unravel the very deepest questions of quantum mechanics and cosmology.

Newton's Absolute Time vs. Einstein's Relativity

Isaac Newton's answer was easy: time simply flows uniformly and without any reference to anything else in the Universe. In Newtonian physics, time is absolute and universal: it is unaffected by motion, gravity, or other forces in the Universe. It serves as the absolute, universal ground that allowed centuries of successful mathematical progress in astronomy, and, of course, led to clear predictions of planetary and other physical motions.

Einstein, however, shattered this notion with his theory of relativity. According to Einstein, time is not an independent quantity but is inextricably linked with space, establishing the four-dimensional framework of spacetime. As we have seen, time can be strained or squeezed depending on the spectator's velocity and the strength of the gravitational field they experience.

The contrast between Newton's absolute time and Einstein's relative time highlights a fundamental shift in our understanding of the universe. For Newton, time was a steady, unchanging river, flowing at an identical speed for everyone. For Einstein, time is more like a flexible, malleable thread, capable of stretching, contracting, and even looping back on itself under certain conditions.

This difference has reflective consequences for our interpretation of reality. While Newton's model provided a solid foundation for classical mechanics, Einstein's framework opened the door to a far more complex and dynamic understanding of the universe. Newton's absolute time was essential for explaining the predictable, clockwork-like behavior of the solar system, but it could not account for the strange and counterintuitive phenomena observed at high speeds or in the presence of massive gravitational fields.

Einstein's theory, by contrast, allows us to understand the behavior of objects and forces at cosmic scales and near the speed of light. It also provides a way to reconcile the behavior of light itself—traveling at a constant speed regardless of the observer's motion—with the behavior of time and space. For example, Newtonian mechanics would predict that if you were moving toward a beam of light, you would measure the light as traveling faster than if you were moving away from it. Einstein's theory, however, shows that the speed of light remains constant for all viewers, despite and regardless to their relative motion. Instead, it is time and space that adjusts to preserve the velocity of light invariant. This fundamental difference between Newton's and Einstein's views of time has real-world implications, particularly in the realm of technology. As mentioned earlier, GPS satellites must account for both the relativistic effects of their velocity and their position in Earth's gravitational field to maintain the system's accuracy. Without these corrections, the GPS system would give increasingly erroneous readings, showing just how critical Einstein's insights are for modern technology.

And yet, both ideas spurred intense discussion about the nature of time over the coming decades: is time an organic feature of the cosmos, or is it the product of space, matter and energy doing something together? Those are still open questions for theoretical physicists today, as they try to unite general relativity with quantum mechanics to create a unified theory.

In conclusion, the shift from Newton's absolute time to Einstein's relative time represents one of the most profound changes in our understanding of the cosmos. While Newton's ideas provided a stable foundation for classical mechanics, Einstein's theory of relativity reveals a universe far stranger and more complex than we could have imagined. As we persist to explore the boundaries of physics, the concepts of time and space will remain central to our efforts to understand the ultimate nature of reality.

CHAPTER 5: THE CONNECTION BETWEEN QUANTUM MECHANICS AND TIME TRAVEL

Quantum Entanglement and Time

Quantum entanglement, which Albert Einstein once dismissively referred to as 'spooky action at a distance', is also incompatible with the standard interpretation of causality as well as time. Causality means that an effect cannot precede a cause, and nothing can travel faster than light. But in the quantum world, phenomena such as entanglement seem to defy these ideas. Particles entangled to one another travel arbitrarily long distances while curiously remaining linked through time and space – if you perform an action on one particle, you can effectively read out the effect it has on the other.

It makes possible the possibility that quantum entanglement might leap not across space but into time as well. There are theoretical physicists who have speculated that, if entanglement can take place between particles across incredible distances, it might also allow for interactions across time. A particle in the present could cause a response in its entangled counterpart in the past or the future. It is a speculation that remains exactly that, but everything changes again if time, this linear arrow slicing through the universe, is really a fluid entity.

Moreover, quantum superposition – another fundamental property of quantum mechanics – also creates difficulties about time. In a state of superposition, a particle has multiple positions or states, until it is measured. So, for instance, a particle might simultaneously be in the state of being in two places, or in two states. It is only upon being measured that it resolves one way or the other. What this means is that the wave function, as it collapses, brings questions like: Does the observation of a quantum event change more than just the present state of the particle? What about the course of its past, leading up to that moment of observation? Are they also altered by the observation? These questions remain at the lead of quantum inquiry.

These quantum ditches become trenches of possibility when applied to time travel. If quantum entanglement allows us to send information instantly across space, then perhaps we can send information – or even causality – across time. In cross-time quantum non-locality, two entangled particles, however separated across space, function as one, as if there is no space between them at all. So, should we suspect that two entangled events or timelines separated across time by hundreds of years could influence one another as well? Perhaps. It is all but impossible to say. But ongoing research into quantum mechanics will continue to challenge our intuitions about time, prolonging its exceptionally radical influence.

The Many-Worlds Interpretation

The Many-Worlds Interpretation (MWI) of quantum mechanics advances a radical solution to the paradoxes often associated with time travel. Projected by physicist Hugh Everett in 1957, this version advances that every outcome of a quantum event occurs in its own branching reality or universe. In other words, the universe is constantly "splitting" into parallel versions of itself, with each split representing a different outcome of every quantum event.

Applied to time travel, the Many-Worlds Interpretation also alleviates most of the paradoxes that haunt the physics of time travel. The old grandfather paradox, for instance: according to the classical views, if a time-traveler goes back and stops himself preventing his grandfather from meeting his grandmother, then he would never have been conceived, meaning he would never have travelled back in time in the first place, a logical impossibility. In the Many-Worlds Interpretation, there is no clear paradox, because the new act of changing the past splits the reality into two new branches. In one of those timelines, your grandfather will never have met your grandmother, but in the other timeline everything is as it was before and there is no contradiction.

The implications of this are profound. If time travel is possible, it may not involve changing the past but rather creating new timelines that diverge from the original. This would mean that every action a time traveler takes spawns a new universe where the consequences of their actions play out, leaving their original timeline unaffected. Time travel would not change the past but create a multitude of parallel futures.

Moreover, MWI raises deep philosophical questions about identity and consciousness. If every decision we make leads to the creation of new versions of ourselves in parallel universes, then what does it mean to be "us"? Are all versions of ourselves equally real, or is the self we experience just one of many possibilities? The Many-Worlds Interpretation forces us to reassess the makeup of personal experience, free will, and reality itself.

Quantum Tunneling and Its Implications

Another way in which quantum mechanics confounds classical common sense is the ability of quantum particles to tunnel outward through a barrier from a region where they are otherwise unlikely to be, a process known as quantum tunnelling. The way this works is that in both classical and quantum mechanics, a particle must have enough energy to jump a barrier. If you want an electron to escape from an atom, for instance, you must give it enough energy to break out of its potential well into a region of lower density of other electrons. But what happens in the quantum world is that an electron, proton, or other particle can also act like a wave. Particles have probability distributions over space, so what happens is that some part of the wave can appear on the other side of the barrier even when classical physics says that this should be impossible.

Quantum tunneling plays a crucial role in processes like nuclear fusion, where it allows particles to overcome the Coulomb barrier and fuse at much lower energies than would otherwise be required. This phenomenon also underpins the operation of many modern electronic devices, such as tunnel diodes and quantum computing systems.

In the context of time travel, some researchers have proposed that if particles can tunnel through spatial barriers, they might also be able to tunnel through temporal barriers. This raises the possibility that quantum tunneling could be used to "jump" through time, bypassing the normal flow of temporal events. While this idea is still highly speculative, it suggests that the same probabilistic rules governing particle behavior in space might also apply to time.

Quantum tunneling also invites us to reconsider the classical boundaries between space, time, and energy. If tunneling can occur in space, could there be a similar effect in time? Could we one day develop technologies that allow us to tunnel through time itself, bypassing the energy constraints that currently make time travel seem impossible?

CHAPTER 6: THEORETICAL MODELS OF TIME TRAVEL

Spacetime Manipulation: Wormholes, Closed Timelike Curves, and Black Holes

Physical models of time-travel usually involve manipulating the very fabric of spacetime. In general relativity, spacetime is one interconnected, pliable entity, warped about its curvature by the presence of mass and energy. Causing spacetime to warp in a certain manner could allow us to open shortcuts through time, even (and this is conjecture) turning its flow backwards so we can 'jump back' in time. Among the most promising types of time-travel are wormholes, closed timelike curves (CTCs) and black holes.

Wormholes, or Einstein-Rosen bridges, are the archetypal theoretical conduit for time travel. A wormhole is a hypothetical tunnel between two widely separated regions of spacetime. If such a conduit could connect two points in space not only in space but also in time, then perhaps it would allow time-travel. The picture of a wormhole is that of a ripple in spacetime, where the spacetime sheet is something that can be bent, waferlike, by gravity.

There are however enormous difficulties – not least that they are dramatically unstable, requiring something called exotic matter (matter with negative energy density) to keep them open, and this has never been observed in the laboratory, in fact its very existence is theoretical. Then there is the problem of the energy needed: to create and stabilize even a human-scale wormhole approaches to what is physically possible.

CTCs are another theoretical model for time travel. A CTC is a closed timelike curve, a loop in spacetime that would allow an object to come back to its own past. The concept of a CTC is a special property of some solutions to Einstein's field equations, most notably in relation to rotating black holes (or Kerr black holes). An object travelling forward in time could theoretically follow a CTC, returning to the point from which it left, and so backward time travel could be initiated.

On the other hand, of course, the presence of CTCs also gives rise to other paradoxes, one of the best-known being the grandfather paradox. As a response to such paradoxes, the Novikov self-consistency principle has been universally accepted. This principle states that, if a story involving time travel would feature a CTC, then this story must be internally self-consistent. That is, the past, which must already exist for the CTC to be possible, already contains whatever the time traveler might do, by design. So, all one could do with time travel is take actions in the past that already have to exist. The time traveler could not do it by changing the past in such a way that it leads to a contradiction.

Black holes, too, are relevant to the subject of time travel, particularly near the event horizons; their strong gravitational fields so skew spacetime that it slows time near the event horizon. If a body is falling into a black hole, then, at least according to the theory, time slows for that falling body with respect to an outside observer. The phenomenon is called gravitational time dilation and has been proved both in theory and observationally. Some theorists believe that certain types of black holes could contain wormholes or other shortcuts through spacetime into other times or other universes.

But time travel using this kind of black hole also presents problems: the intense gravity of black holes will destroy any object that tries to cross the event horizon, and the black hole information paradox – the apparent loss of information about the physics of an object's state when it falls into a black hole – is a mystery we have yet to solve. A different solution to the time travel problem is to try 'wormholes': rather than passing through the same region of our fabric of space and time, you head through some other region of space, popping out into the past. Taking this approach complicates an already complicated matter. Stepping into a wormhole means travelling at fast-than-light speeds.

In sum, while these models remain speculative, they do extend the frontiers of spacetime and offer interesting directions for future research. Perhaps one day, when we have a deeper and more unified understanding of quantum mechanics and general relativity, we will know if these theoretical avenues can be realized.

CHAPTER 7: PRACTICAL CONSIDERATIONS AND CHALLENGES

Energy Requirements for Time Travel

One of the most substantial contests to turn the time travel a reality is the colossal energy needed to affect spacetime. Theoretical models such as wormholes and cosmic strings suggest that traversing time would demand energy levels far beyond what we can currently generate. For example, stabilizing a wormhole large enough for human travel would entail exotic matter with negative energy density, a type of matter that has yet to be discovered. Even if such matter were found, the energy required to maintain the wormhole would be comparable to the mass-energy of entire stars.

Another theoretical model, cosmic strings, are one-dimensional defects in spacetime that might have modeled in the early universe. Two cosmic strings, if brought close enough together, could hypothetically generate a time loop and allow backward time travel, according to some physicists. Unfortunately, attempting to manipulate cosmic strings would require energies so high that they are nearly incomprehensible at the present time: any existing structures like this would be extremely dense and difficult to manage.

In addition to these models, quantum mechanics introduces its own set of challenges. Phenomena like quantum superposition and entanglement suggest that time travel might not require physically moving through time but instead manipulating quantum states. However, the energy required to harness these effects would still be astronomical. Moreover, the unpredictable nature of quantum mechanics adds another layer of complexity, as controlling quantum states over extended periods is exceedingly difficult.

While the dream of time travel may be alluring, the energy

requirements present formidable obstacles. Theoretical models provide fascinating frameworks, but they also underscore the vast gulf between theory and practical reality. As researchers continue to explore the depths of quantum mechanics and general relativity, we may one day unlock the secrets of time travel, but for now, it remains a distant possibility.

The Paradoxes of Time Travel

The notorious time-travel paradoxes show us that our intuitions about causation leap off the rails as soon as we allow for time travel. The most celebrated of these is the grandfather paradox – a situation in which a traveler might go back in time, thereby changing what happened in that past, and in the process preventing his or her own birth. But if the traveler's actions lead to the traveler never being born, how could he or she have travelled back in time in the first place? The grandfather paradox vividly argues against going back in time. If the traveler prevents his or her own existence, they could not have gone back in the first place. It seems that backward time travel is impossible in a single, linear timeline.

Another is the bootstrap paradox, in which something or somebody sent back in time originates in the future; for instance, if a time-traveler gives a copy of Shakespeare's works to Shakespeare, who wrote the works? The bootstrap closes time down to a time-loop, an eternal cycling of events, where it is impossible to tell where time begins.

These paradoxes imply that causality is logically compromised by time travel, and that the possibility of time travel needs to be parsed differently when using frameworks such as the MWI. In this model, travelling back in time does not change the original time-line, but creates (actually, selects) new wrinkled versions of the time-line where the consequences of the traveler's action play out independently. This respects causality – but at a meaningful cost to our conventional understanding of reality, and who we are.

CHAPTER 8: PHILOSOPHICAL IMPLICATIONS OF TIME TRAVEL

The Nature of Free Will

The question concerns free will and how its relationship to time travel is understood in the context of quantum mechanics. In classical physics, the future happens deterministically – if we knew everything about the present, we could predict the future with certainty. But quantum mechanics has stochastic elements that seem to indicate that not everything is predetermined. This gives rise to the possibility of free will, with individuals choosing events that would not have been determined by the preceding ones.

Free will issues are even more complicated in the context of time travel, especially if it is possible for a being to travel to the past and alter the course of events – is that a choice, or an instance of creating a new time-line? The Many-Worlds Interpretation permits both outcomes to exist simultaneously, making it possible for you to really have a choice without disrupting the timeline. If that timeline includes every potential outcome, then at what point does free even enter the picture? If the past and future are already set in stone, then isn't one's inevitability simply the result of ineluctable processes?

Ethics and Morality in Time Travel

Time travel also introduces significant ethical dilemmas. The ability to modify historical events builds questions about responsibility and the consequences of such actions. Should time travelers intervene to prevent tragedies, such as wars or natural disasters, or does such interference risk creating unforeseen consequences? The butterfly effect, where slight changes in the past can have large impacts on the present and future, underscores the potential dangers of time travel.

Moreover, time travel could be manipulated for personal benefit and profit. If someone had knowledge of the future, they could manipulate markets, alter historical outcomes, or erase competitors from existence. Such actions raise concerns about fairness and justice, highlighting the need for ethical frameworks to guide decisions in a world where time travel is possible.

Time Travel in Historical Context

The concept of time travel has deep roots in literature and philosophy, long before it became a subject of serious scientific inquiry. Early myths and stories often featured heroes who could travel between separate times, challenging the linear perception of time that dominates Western thought. However, it was not until the 20th century, with the development of Einstein's theory of relativity, that time travel gained scientific legitimacy.

Relativity showed that time is not an absolute entity but a flexible element or dimension that can be warped by mass and energy. This revelation opened the door to theoretical models of time travel, such as wormholes and closed timelike curves. As our understanding of spacetime has advanced, these models have shifted from the scope of science fiction to serious scientific exploration.

Quantum mechanics has further complicated the picture, introducing concepts like superposition and entanglement that challenge conventional understandings of time and causality. While practical time travel remains a distant possibility, the exploration of these principles continues to push the boundaries of what might be possible, inviting deeper inquiries into the nature of time and our place within it.

CHAPTER 9: FUTURE OF TIME TRAVEL RESEARCH

Current Theories and Experiments

The scientific and popular discourse on time travel has grown from scientific speculation by Albert Einstein, subsequently sustained by modern quantum physics, to a subject of serious scientific investigation. Current research on time travel is based on the complex quantum mechanical and relativistic conceptions. Although at present the subject of mere speculations, the avenues it opens for a future investigation are too vast to be ignored. Perhaps the search could throw some light on the working of a mechanism that makes time-travel possible — someday.

One up-and-coming theoretical tool in this area is the wormhole, which is defined as a contemplated shortcut through spacetime, linking together distant points in space and time. A special case of the so-called Lorentzian manifolds which result from general relativistic spacetimes, the generic wormhole has been theorized since the invention of Einstein's general theory of relativity in 1915 (sometimes consequently labelled as an 'Einstein-Rosen bridge'). Wormholes mediating faster-than-light travel might link arbitrarily distant locations in the Universe, but by selecting pairs of such distant points (if this is even possible), one could exploit wormhole theory to find the calculable connection via a spacetime channel that could potentially allow one to circumnavigate the time gap. Wormholes remain, however, very speculative steeds of time travel, since the stability problem of creation is insurmountable, at the moment. For a wormhole to maintain a two-legged permanent open channel linking the two ends, it will (according to the current state of the art) need negative energy exotic matter as a prop, which remains an open problem. Still, wormholes remain a cornerstone of advanced theoretical physics.

Among other time-travel theoretical constructs, another conspicuous one is closed time-like curves – spacetime loops that would (theoretically) allow an object to travel back to its own past, and a whole host of resulting paradoxes, most famously the 'grandfather paradox' (what happens if a time traveler prevents her grandfather from meeting her grandmother?). Though time travel along CTCs is possible because of general relativity, these have fantastic implications for causality and free will, and philosophers and mathematicians have since developed models under which CTCs can exist despite known physics laws, as well as many other thought experiments concerning paradoxes.

We are also making progress through experimental approaches that complement these theoretical frameworks. In the past few decades, we have seen some promising new directions emerge in our understanding of time and how we might one day bend it to our will. One of the most compelling areas relates to physics at the quantum scale. We can learn a lot from the study of particles moving at extremely high speeds, where we have had the opportunity to observe the phenomenon of time dilation firsthand by witnessing how such particles perceive the flow of time when placed inside particle accelerators like the Large Hadron Collider (LHC) at CERN in Switzerland. As per Einstein's general theory of relativity, an object moving at speeds close to the speed of light will speed up relatively to an observer who is standing still, with the passage of time itself apparently slowing down accordingly. Although this does not translate into time travel (yet), the fact that we can observe theoretical effects of time manipulation experimentally is an encouraging sign, and it is a whole lot closer to our current understanding of the physicochemical world. Likewise, we are also learning more about the malleability of time by studying the effects of gravity.

Quantum effects are another subject of interest. Unlike entanglement, quantum effects are not readily constrained by our understanding of classical physics. One quantum effect that is the subject of intense interest within the time-travel community is the concept of entanglement. Two particles that are entangled interact with each other in a unique way: the state of one instantaneously influences the other, no matter how long the distance that separates them. Bell's Theorem quantifies how surprising this behavior is, but not how surprising it ought to be to justify the idea that we could use it to send information backwards in time. However, entanglement is startlingly wild. It amounts to a quantum connection between two particles that deeply disrupts classical ideas about time and space. It is natural to think that there might be a more straightforward explanation, but current research has not produced one. One way might be to generalize the idea of 'instant' connection in entanglement to allow for transmitting information between times. No one could tell us how to do this, and no one has presented even a speculative theory for how reverse temporal radio could have been evaluated in the lab, but it is true that quantum physics distinguishes between far and near in strange ways.

Overall, I think we can say that these ideas about time travel, whether they involve wormholes or closed time-like curves, or involve quantum mechanics, are an exciting frontier of scientific research. We cannot yet say whether time travel is possible, but perhaps one day, if physicists discover the true, underlying nature of time and space, then those discoveries might lead to more options for how we might move – or 'move' – through the flow of time. It might be that time is more complicated and dynamic than we have known. But for now, I am looking forward to our next physics talk.

Potential Breakthroughs in Quantum Physics

Quantum physics is one of the most promising fields in the study of time travel. It challenges the very foundations of how we understand reality and offers potential pathways to manipulating time itself. In recent years, advancements in quantum mechanics have revealed fascinating phenomena such as superposition, quantum tunneling, and entanglement, all of which could hold the key to understanding how time travel might be realized.

One of the most encouraging portions of quantum physics is quantum entanglement. As previously discussed, entangled particles can affect one another promptly, even when split by enormous gaps. This phenomenon violates the classical understanding of time and causality, as information is transferred faster than the speed of light. If entangled particles could influence each other not only across space but also across time, it could provide a mechanism for time travel. Current research is exploring whether entanglement could be harnessed to transmit information into the past or future, though such applications remain theoretical at this stage.

Superposition is another quantum phenomenon that has effects on time travel. In the quantum world, particles can exist in multiple states simultaneously until they are observed. This "probabilistic" nature of quantum mechanics suggests that time may not be a simple linear progression but rather a complex tapestry of potential realities. If scientists could find a way to manipulate these quantum states, it could theoretically open the door to time travel by allowing access to different points in time. While we are far from achieving this level of control, the study of superposition continues to push the boundaries of what is possible in terms of time manipulation.

Quantum tunnelling is another way that has been discussed as a possible clock mechanism for time travel. In classical physics, particles need a certain energy level to overcome an energy barrier. However, the peculiarities of the quantum world indicate that particles can 'tunnel' through barriers they should not be able to pass, according to classical models. Since classical mechanics says you cannot violate certain laws of physics, some researchers have speculated that particles could also 'tunnel' through time. This is still very speculative, but it shows that the weird nature of quantum mechanics is changing the rules that have guided previous thinking about the way the Universe works.

As these quantum physics breakthroughs keep coming, they prompt deeper philosophical questions, too. What would it mean for the idea of 'destiny' if time-travel were definitively possible? If time-travel were possible, would you be able to change your past, or would you find you were subject to self-consistency principles that avoid paradoxes? These are all highly theoretical questions, but they are at the forefront of both scientific and philosophical discussion of the possible results of manipulating time.

While quantum physics is still in its initial stages concerning time travel, its recent advancements suggest that the future of time exploration may lie in the quantum realm. Whether through entanglement, superposition, or quantum tunneling, scientists are increasingly exploring how the quantum world could provide new avenues for understanding and manipulating time.

The Future of Human Exploration of Time

Human fascination with time travel is deeply ingrained in our culture. From early mythology to modern science fiction, the perception of traveling to the past or future has captivated imaginations for centuries. As our understanding of quantum mechanics and relativity grows, what was once the domain of fantasy is slowly transitioning into the realm of scientific possibility. The future of human exploration of time is poised to be one of the most exciting frontiers of scientific discovery.

Quantum mechanics, as discussed earlier, offers some of the most promising avenues for time exploration. If we can harness phenomena like entanglement or superposition, it may be possible to manipulate time in ways previously thought impossible. This could include sending information about the past or even physically traveling through time. While the technology required to achieve this level of control is still far beyond our reach, the groundwork is being laid by researchers in quantum physics.

But it also relies on relativity; Einstein's theories of gravity and of high speed indicate that time can bend and even stretch and contract. Time dilation has been demonstrated experimentally, as when a clock ticks more slowly for a moving object than for a stationary observer when positioned at the destination. It is not really time travel but the relativity principle does suggest human travelers might one day experience time differently, through relativistic or very long journeys – or as a wonderful sidestep, through a variety of other methods we have not imagined yet.

There are the cultural implications, too. Stories involving time travel have been part of literature, film, and other media for ages; and have begun to shape perception about what time travel would mean, if it was real. As this science moves closer to becoming reality, we will begin to have real questions about what should be allowed, what would be permissible – and what the consequences of tinkering with time would be. Would messing with the past be a good thing? Or a bad one? How might that impact the present – and future – in turn?

In the end, it seems likely that human exploration of the realm of time will unfold within these two paradigms: quantum mechanics and relativity. Quantum travel might seem the easier of these options, because deliberately engineering quantum phenomena is something that we are already getting to grips with; quantum entanglement already exists in every laser pulse, after all. But there are plenty of experimental challenges to be overcome before engineers can decide to send a box of entangled particles back in time. Relativistic travel, by contrast, has sown the seeds of its own experimental humiliation – do you really want to pick up speeds so disobedient to matter, from the point of view of an inside observer, that they make you appear to blast apart? Something tells me you can count me out on that proposition. That is not to say, however, that I have altogether given up on the prospect of time travel. Quite the opposite. I submit that the possibility of time travel, at the very least, is one of the most fascinating frontiers of the contemporary scientific imagination. Very possibly, it is the most fascinating. The road ahead is full of thorns, yet it is also filled with promise: of new machinery, of unlocking the wellsprings of our physical world, and of learning how to shape reality itself.

CHAPTER 10: THE MULTIVERSE THEORY: PARALLEL UNIVERSES AND TIME TRAVEL

The Concept of Multiple Universes

The idea is that we live in one universe out of an infinite number of universes – an infinite multiverse. Everything happens in our universe, and rather than everything being busy at the same time, it is busy at different points in time across the entire spectrum and range of possibilities. But does any of it really happen in a multiversal sense? Are we, as quantum mechanics suggest, face to face with a picture of everlasting possibilities? Is all that alone testament to the feasibility of preempting the passage of time by way of pure intellectual prod? It is difficult to know. But if you believe a glimmer of truth lies within theories of time travel, then behold after seventy years of silence, the Daedalus Project has been given new life.

The most potent defense of the existence of multiple universes comes from the Many-Worlds Interpretation (MWI) of quantum mechanics – or 'quantum weirdness'. On this view of the world, every quantum event is interpreted as a split in the Universe into a fork (many worlds'). There is a universe in which you take the left door, and there is a universe in which you take the right one. And when you or I go through that door, the Universe splits at that instant over and over, as we do a million different things each second. The result? Almost infinite parallel realities.

Then again, the Many-Worlds Interpretation, which proposes that the wave function splits into a multiplicity of parallel universes every time a real decision must be made, might provide an escape from those paradoxes – at least for time-travelling scenarios that alter the past. With each action, the Many-Worlds Interpretation suggests, the reality-splitting into multiple timelines signifies the creation of a new universe, rather than the 'change' of an original one. So, the trip back in time, to prevent killing your grandfather, for example (the classic grandfather paradox), would not alter any past events, but merely prevent you from ever being born at all, by making sure your parents never met. It could be as simple as that. Yet another window into the multiverse comes by way of cosmological theories, one of the most influential being the theory of cosmic inflation. During this early phase of the universe's development after the Big Bang, which could have lasted for less than a year, the expanding fabric of space distributed bits of energy in a completely homogenous way, like sugar dissolved in a cup of tea. This made for an even distribution of energy within every region of space. Gravity caused the regions of space to shrink, cooling them to such a degree that they could be filled with particles, which eventually condensed into matter and were pulled by gravity to form primordial galaxies, and ultimately, to forge our universe out of others. The final stage of inflation, which spawned a multiverse's worth of cosmological bubbles, could have originated in space near our universe, which would be the only one whose bubbles ultimately stayed inflated. Or it could have sprouted from other regions of space, which might have cooled off faster than ours, evolving as individual universes neither like nor unlike our own. In this picture, a multiverse would form an infinite cosmic landscape where your universe is but one bubble in a vast sea of many.

General relativity itself provides potential avenues for crossing between the universes: the warping effects of spacetime curvature, the pulling of massive objects like black holes, and the concept of wormholes connecting otherwise distant points in time and space would provide the means to not only travel through time in our own universe, but to traverse between universes as well. All of this remains speculative, of course, but it shows how the theory of relativity, by expanding our understanding of space and time, would open the possibility of time travel into the multiverse.

Indeed, the philosophical implications of multiverse theory are potentially significant. If every possible outcome is realized somewhere in a universe, then the question of what happens certainly has an answer.

Quantum Superposition and Entanglement: Their Role in Theoretical Time Travel

These are two of the fundamental phenomena of quantum mechanics that have made the subject so difficult to accept. The quantum phenomenon of superposition allows particles to exist in more than one state at the same time. The phenomenon of quantum entanglement connects pairs of particles such that, if either is measured to be in a specific state, the state of the other immediately establishes a perfect correlation with that state, even if the two particles are widely separated. These two phenomena together enable us to develop a reasonable speculative account of how time-travel might work.

If particles can exist in a superposed state, why not time itself? The traditional model of time as a fixed coordinate assumes that superposition cannot apply to time, but there is no logical reason to dismiss the possibility. A 'leapfrog' model of time could allow it to unfold in superposed states. From this perspective, traveling back and forth in time, like a quantum spacetime jump, could simply involve stepping across quantum states in a multidimensional time framework.

Entanglement also muddles our expectations about time: if the states of two particles are entangled, the states are linked despite huge interparticle distances, and a change of the state of one particle will affect the other instantaneously. Could it be that time and space are more tightly connected than we think? If particles can be entangled across time, perhaps information (or physical matter) could be sent back in time. This would provide a mechanism for time travel that does not require the classic form of travel into the future in ordinary spacetime.

Indeed, similar rules could potentially apply to quantum tunnelling, where particles are known to 'tunnel' through barriers that they would not necessarily pass through in the real world; if particles can tunnel through space, why not through time? While much of this is speculation, it does point to a new approach to thinking about time-travel, one that sticks much more closely to the counterintuitive and downright weird world of quantum mechanics than some previous formulations.

Those same concepts, as researchers continue to explore quantum superposition and entanglement, are bound to play an ever more emphatic role in those predictive imaginings. It may seem remote, but time travel will someday be quantum. Meanwhile, all one can do is marvel at how this miraculous reality we call the quantum world might very well transform concepts such as causality and, in line with that, the notion of time's flow. The concept of time travel might also manifest itself in unexpected ways, as physicist Adrian Kent from the University of Cambridge muses, or it might not. But it will, at the very least, ultimately, tell us the way back, from physics to ourselves.

CHAPTER 11: THE GRANDFATHER PARADOX AND OTHER TIME TRAVEL DILEMMAS

Time Travel Paradoxes: The Grandfather and Bootstrap Paradoxes

The fantastic, zany, and futuristic aspects of time travel are balanced by some serious and perplexing dilemmas that time travel raises. If it were true, what would time travel do to causality and our understanding of the nature of time itself? Perhaps the best-known example of this set of puzzles is the Grandfather Paradox. It begins with an assumption: Suppose that a person time-travelled into the past, and killed her grandfather before he met her grandmother, and so she would never have been conceived. Then she would not time-travel to the past and could not have killed her grandfather. Contradiction! How could she have travelled to the past in the first place?

The **Grandfather Paradox** challenges how we understand time and causality. If you went back in time to kill your grandfather, it would prevent your own birth, leading to a contradiction. In some interpretations, such as the **Many-Worlds Theory**, changing the past might result in branching timelines, where a new universe is created, leaving the original timeline intact. Instead of rewriting history, these "branching worlds" allow multiple versions of reality to coexist, potentially resolving the paradox by avoiding direct contradictions.

A more interesting paradox of time travel is the Bootstrap Paradox that deals with the issue of self-generation. Greatest-hits questions for the Bootstrap Paradox pose the conundrum of a physical object or information that is sent back in time and causes itself. A time-traveler might, for example, return to the past and hand-deliver a copy of Shakespeare's collected works to William Shakespeare himself. Shakespeare publishes these plays – but who wrote them? If the plays could not exist until they were sent back from the future by a time-traveler, then, the paradox tells us, there is no point at which they come into being.

What emerges from each of these time-travel paradoxes is a potential threat to the causal structure of time – the deeply held belief we have that causes happen before their effects. In each case, the story we tell ourselves about time turns out to be more intricate and trickier than was first apparent – which is not necessarily a problem. But it is intriguing to note just how uncomfortably these Grandfather and Bootstrap Paradoxes press against our intuitions about time and causality, and how each of them in their own way seems to hint that travelling back in time could have wide-ranging and confusingly profound effects on the fabric of the Universe. These thought experiments about termination continue to serve as a rich source of new tensions and dilemmas in the ongoing philosophical and scientific enquiry into the nature of time.

Implications for Time Travel and Reality

The implications of time travel paradoxes extend beyond theoretical discussions; they have profound inferences for our interpretation of reality itself. If time travel were possible, how would it affect the stability of the universe? Could time travelers alter momentous events in history, leading to unpredictable changes in the present or future? These questions confront our understanding of causativeness and indicate that time travel might not be as simple as moving from one point in time to another.

Some physicists propose that time travel might be possible through mechanisms like wormholes or closed time-like curves. However, the existence of paradoxes raises concerns about the stability of these models. If time travelers can alter the past, it could lead to chaotic outcomes that defy our understanding of physics. These concerns highlight the need for further research into the description of time and the prospective consequences of time travel.

In conclusion, time travel paradoxes like the Grandfather and

Bootstrap Paradoxes serve as thought experiments that challenge our perceptions of time, causality, and reality. They push us to encounter the restrictions of our understanding and the potential complexities of traveling through time. While these paradoxes may seem purely theoretical, they provoke meaningful discussions about the fabric of the universe and our place within it.

CHAPTER 12: ELEVEN DIMENSIONS OF REALITY

The Concept of Higher Dimensions

In the domain of quantum mechanics and theoretical physics, the notion of dimensions reaches far beyond the three-dimensional world we encounter daily. Traditional physics describes a universe made up of three spatial dimensions—length, width, and height, and one temporal dimension: time. However, contemporary theories, particularly string theory, propose that our universe may encompass up to eleven dimensions. These additional dimensions offer intriguing prospects for interpretation of the nature of reality and time itself.

The first three dimensions allow us to describe the physical properties of objects and their positions in space. The fourth dimension, time, adds a crucial layer to our understanding of how events unfold and interact. Einstein's theory of relativity fundamentally modified our view of time by suggesting that it is not separate from space but is intertwined with the three spatial dimensions, forming what is known as spacetime.

As we venture beyond the familiar four dimensions, we enter realms that challenge our understanding of existence. The fifth dimension introduces the concept of alternate realities or parallel universes. Here, every decision or event might lead to a branching path, creating multiple futures. This concept finds support in certain interpretations of quantum mechanics, where particles exist in a superposition of states until observed. The fifth dimension, therefore, represents a richer tapestry of reality where countless variations coexist simultaneously.

The sixth dimension and the seventh, however, are even more conceptual and abstract. The sixth dimension represents a plane of possibilities: for every actual outcome of an event, there is another branch of the Universe where that event manifests as a different possibility. Thus, not only is there a parallel Universe for everything that could have happened, but for everything that could possibly have happened. The seventh dimension then takes this one step further: each iteration also uses different laws of physics entirely. Entire universes could be made up of rules entirely incomprehensible to us, showing that there is no limit to what reality might do in principle.

The eighth through eleventh dimensions continue this exploration into the unknown. The eighth dimension might allow for the manipulation of time itself, offering pathways to influence past events. The ninth dimension could represent a space where all laws of physics converge, while the tenth dimension encapsulates the entirety of every universe and timeline. Finally, the eleventh dimension is theorized to be a realm of pure potential, where all dimensions converge into a unified whole. While these concepts remain speculative, they invite us to imagine a universe far more complex and intricate than previously thought.

The 11 Dimensions from String Theory and Their Relationship to Time

String theory is one of the most ambitious tries to unite our recognizing of the universe. It aims that the structural building blocks of certainty are not particles but tiny vibrating strings that exist in higher-dimensional space. According to string theory, our universe is composed of eleven dimensions, most of which are hidden from our everyday perception.

The first four dimensions—three spatial and one temporal—are familiar to us. However, the additional dimensions described by string theory are compactified, meaning they are folded in on themselves and exist at incredibly small scales. These hidden dimensions could help clarify some of the most perplexing mysteries of the universe, including the makeup of time.

In string theory, the extra dimensions may provide a framework for understanding how time behaves on both macroscopic and quantum scales. The fifth dimension, for example, could be related to the Many-Worlds Interpretation of quantum mechanics, where different outcomes of quantum events create branching realities. The sixth dimension might represent the full range of possible histories and futures for every particle in the universe.

As we advance into the higher dimensions, the possibilities become even more abstract. The seventh through eleventh dimensions could offer new ways of thinking about time travel and the manipulation of spacetime. While these ideas remain speculative, they represent an exciting leading edge in our construction of the universe.

In conclusion, the concept of eleven dimensions from string theory provides a fascinating context for understanding the nature of time and reality. By exploring these higher dimensions, we gain new understandings into the underlying processes of the universe and the potential for time travel. Although we are far from fully understanding these concepts, the pursuit of knowledge in this field promises to revolutionize our understanding of existence and open new doors to the mysteries of time.

Chapter 13: Time Travel in Quantum Gravity and String Theory

Quantum gravity and string theory are among the most magnificent theoretical frameworks in modern physics, aiming to answer some of the most extreme questions about the universe. These theories are particularly intriguing because they may offer insights into one of humanity's most exciting and perplexing possibilities: time travel. By bridging quantum mechanics and general relativity, both theories explore fundamental aspects of reality that could make time travel feasible.

At the kernel of these theories is the endeavor to reunite quantum mechanics, which runs the behavior of particles at the smallest scales, with the force of gravity, which general relativity explains as the curvature of spacetime. These two domains of physics are famously incompatible, and yet both are necessary to understand the universe fully. Time travel is a concept that sits at the intersection of these two theories, requiring a unified framework to navigate the quantum world and the gravitational forces shaping spacetime.

Quantum gravity holds potential for resolving paradoxes that arise in classical physics, such as those associated with time travel. In classical mechanics, traveling back in time introduces problems like the "grandfather paradox," where modifying the past could prevent one's own actuality. However, quantum gravity suggests that time might not follow the linear, immutable progression we typically assume. Instead, it might possess a more complex, malleable structure, with multiple timelines or parallel universes. In these scenarios, changes to past events would create new branches or realities, preventing paradoxes from undermining causality.

String theory, another major theoretical contender, builds on the concept that the universe's fundamental elements are not point-like particles but miniature, vibrating strings. This model introduces further dimensions beyond the familiar three of space and one of time, proposing that the universe may contain as many as 10 or 11 dimensions. Some researchers suggest these extra dimensions could contain wormholes—shortcuts through spacetime that could allow for travel between distant points in time. If wormholes exist, they could enable time travelers to leap between the past, present, and future, much like stepping through a tunnel between various parts of spacetime.

These ideas contest our perception of time as a fixed, linear dimension. According to both quantum gravity and string theory, time could be manipulated under specific conditions, such as by warping gravitational fields or stabilizing wormholes. Although highly speculative, these concepts inspire innovative approaches to time travel, pushing the boundaries of what we perceive as possible in both science fiction and theoretical physics.

In summary, quantum gravity and string theory represent groundbreaking frameworks for rethinking time travel. By offering new perspectives on the nature of time, space, and the fabric of the universe, these theories might one day provide the answers needed to turn time travel from a theoretical curiosity into a scientific reality. As research into these fields continues, we move closer to unraveling the mysteries of time itself and potentially discovering how to traverse it.

How Advanced Mathematical Theories Provide Speculative Paths to Traveling Through Time

Mathematical theories are key to understanding time travel, with quantum mechanics and general relativity at the core. General relativity, Einstein's theory, introduces spacetime as a flexible fabric shaped by mass and energy, predicting wormholes—hypothetical tunnels connecting distant points in space and time. If stabilized, wormholes could enable time travel. The theory also suggests closed timelike curves, which allow for backward and forward movement in time, though they raise paradoxes like the grandfather paradox. Some physicists argue quantum effects may resolve these issues by splitting timelines into multiverses.

Quantum mechanics adds complexity with superposition, where particles can exist in multiple states, and entanglement, which connects particles across vast distances. These phenomena hint at alternate timelines or parallel realities, offering new possibilities for time travel. Though practical time travel remains speculative, these mathematical models inspire ongoing scientific exploration, suggesting that time travel may one day be more than just theoretical.

CHAPTER 14: ETHICAL IMPLICATIONS OF TIME TRAVEL: SHOULD WE DO IT?

The Social Consequences of Time Travel

The prospect of time travel has fascinated humanity for centuries, but its potential social consequences have often been overlooked in popular discussions. If time travel were to become an actuality, it would have extreme effects for society, fundamentally altering our relationships, understanding of history, and ethical frameworks. The ability to move through time introduces complexities that go far beyond the technical challenges, raising questions about how such power should be wielded and by whom. One of the most abrupt social consequences of time travel would be its impact on interpersonal relationships. Imagine the emotional and psychological complexities that could arise from the ability to visit loved ones in the past or future. While this might deepen connections with people we have lost, it could also lead to confusion, guilt, or even resentment. For example, a person might travel back to witness a significant event in their life, only to return with a changed perspective on their relationships. Similarly, the ability to interact with past or future versions of oneself could create an intricate web of conflicting loyalties, desires, and regrets, complicating the very essence of human interaction.

Time travel could also dramatically redesign our perception of history. The ability to observe historical events firsthand would challenge traditional narratives and potentially disrupt the education system. Scholars would have to reconsider how history is taught, as first-hand experiences might contradict established facts. While this shift could lead to a more nuanced understanding of historical events, it could also foster skepticism toward the reliability of historical accounts. People may begin to question the authenticity of events, as the possibility of altering the past through time travel could cast doubt on what truly happened.

Beyond personal and historical implications, time travel raises significant ethical dilemmas. The ability to change past events forces us to confront questions about free will and moral responsibility. If an individual could prevent a tragedy or alter a significant outcome, what ethical obligations do they have? The potential to create alternate realities or timelines introduces further complications. Would time travelers bear responsibility for the consequences of their actions in these alternate timelines? The ripple effects of altering even a small event in the past could have far-reaching consequences, raising questions about whether anyone should have the right to make such changes.

The societal structure itself could be transformed by the advent of time travel. Economically, the ability to access past resources or knowledge could disrupt markets and industries. For instance, individuals who travel to the future might gain insights into stock market trends, technological innovations, or political developments, giving them an unfair advantage. This could lead to economic instability and exacerbate existing social inequalities. Moreover, conflicts could arise as distinct groups vie for control over time travel technology, leading to power struggles that could reshape societal hierarchies. Governments and international organizations need to establish strict regulations to inhibit the abuse of time travel and to ensure that it does not fall into the wrong hands.

If Time Travel Were Feasible, Should Humanity Embark on this Journey?

The question of whether humanity should pursue time travel goes beyond mere scientific curiosity or escapism. The ability to alter historical events or visit the future presents profound ethical, social, and philosophical challenges. Should we use this power to prevent past tragedies, or would such actions have unintended consequences that could spiral out of control?

One of the primary concerns surrounding time travel is the potential for changing historical occurrences. The idea that one could go back in time and change a pivotal moment raises intricate ethical dilemmas. For instance, if someone were to prevent a historical tragedy, would this act create a ripple effect that alters the course of history in unforeseen ways? This notion is famously illustrated in the "butterfly effect," where even small, insignificant changes can lead to significant consequences. Without a comprehensive framework to govern time travel, humanity could inadvertently jeopardize its own timeline.

Furthermore, the pursuit of time travel presents significant scientific challenges. While Einstein's theory of relativity suggests that time is a dimension that can be manipulated under specific conditions, creating a mechanism for time travel would require technological advancements and an understanding of physics that currently elude us. The resources required for such research could divert attention from more immediate global challenges, such as climate change, poverty, and healthcare. While time travel might yield benefits in various fields, such as medicine and energy, the question remains: is it worth the cost?

On a societal level, time travel could exacerbate existing inequalities. If access to time travel technology were limited to the wealthy or privileged, it could create a new class of time travelers who wield disproportionate power over the past and future. This could lead to conflicts over who has the right to alter history and who bears responsibility for the consequences of such actions. Society needs to adapt to these changes, raising questions about governance, regulation, and the moral responsibilities of time travelers.

In conclusion, while time travel offers enticing possibilities, humanity must weigh the potential benefits against the ethical, social, and scientific challenges it presents. The allure of revisiting the past or exploring the future is undeniable, but the consequences of our actions could be far-reaching and unpredictable. Before embarking on this journey, we must thoroughly consider whether the hazards outweigh the incentives.

CHAPTER 15: TIME TRAVEL AND SLEEP: THE DREAM OF TIMELESS JOURNEYS

Time Travel Within Dreams

Sleep has long been regarded as a mysterious bridge between the conscious and unconscious realms, with dreams offering a window into other timelines or realities. Many cultures throughout history have ascribed spiritual or prophetic significance to dreams, believing that they hold the power to reveal hidden truths or glimpses of the future. From this perspective, sleep may be the closest experience we have to time travel, allowing us to explore the past, present, and future without physically moving through time.

Dreams often distort our perception of time, creating the sensation that hours have passed in a matter of minutes, or compressing lengthy scenarios into brief moments. This temporal fluidity mirrors the concept of time travel, where memories of the past and projections of the future blend together in a seamless flow. In this sense, our nightly dreams can be seen as a form of mental time travel, where we slip through the folds of time without ever leaving our beds.

The view of time travel within dreams has captivated scientists and philosophers for epochs. Some have speculated that dreams allow us to access different dimensions or timelines, where we can relive past experiences or catch glimpses of potential futures. While this idea remains speculative, it raises interesting questions about the nature of time and how our minds perceive it during sleep. Could our dreams be more than just figments of our imagination? Could they serve as a gateway to other realms of time and space?

The Science of Dreams and Time Perception

Research into the nature of dreams and their relationship to time perception has revealed fascinating insights into how our brains process time while we sleep. Most of our dreaming occurs during Rapid Eye Movement (REM) sleep, a phase characterized by vivid, often bizarre dreams. During REM sleep, the brain is substantially active, and our perception of time can become distorted. For example, we may experience long, detailed dreams in a short span of time, or conversely, we may wake up feeling as though only a few minutes have passed, even though we have been asleep for hours.

One rationalization for this trend lies in the brain's altered state of consciousness during sleep. The prefrontal cortex, the part of the brain responsible for logical reasoning and time perception, becomes less active during REM sleep. This allows the dream state to blend different timelines, creating the sensation that past, present, and future events are occurring simultaneously. In this state, our minds are free to explore different moments in time, unhindered by the linear constraints of waking life.

Neuroscientists have also studied the relationship between sleep and time perception, particularly in the context of lucid dreaming in which the dreamer becomes informed that they are dreaming and can sometimes control the substance of their dreams. In lucid dreams, individuals often report experiencing time in ways that defy logic, such as reliving entire days in a single dream or traveling to different points in their lives. While the scientific understanding of these phenomena is still in its preliminary stages, they offer intriguing possibilities for how time might be experienced differently in the dream state.

Historical Perspectives: Sleep as a Portal

Throughout history, sleep has been viewed as a portal to other dimensions and realms of existence. In archaic civilizations, dreams were habitually perceived as a way for the soul to leave the body and travel freely through time and space. For example, the ancient Greeks alleged that dreams could carry messages from the gods, offering glimpses of future events or insights into past experiences. Similarly, in many indigenous cultures, shamans used dreams as a tool for spiritual journeying, traveling to other worlds or dimensions to gain wisdom and guidance.

In literature, the theme of sleep as a gateway to time travel has been explored in numerous works. H.G. Wells' classic novel *The Time Machine* describes the protagonist's journey through time as a kind of dream-like experience, where time compresses and expands beyond his control. Similarly, Mark Twain's *A Connecticut Yankee in King Arthur's Court* begins with the protagonist falling asleep and waking up in a different century, blurring the lines between sleep and time travel. These stories reflect a deep-seated fascination with the idea that sleep could serve as a bridge between different moments in time.

Sleep for Long Interplanetary Trips

As humanity looks toward the stars and contemplates the possibility of long-duration space missions, the concept of sleep as a tool for time travel takes on new significance. Traveling to Mars, for example, would take several months to years, depending on the alignment of the planets and the propulsion expertise applied. For astronauts, spending such a long time in the confined space of a spacecraft presents numerous physical and psychological challenges.

One potential solution that scientists are actively researching is the use of sleep-like states, such as torpor or suspended animation, to help astronauts endure these long journeys. Torpor is a state in which an organism's metabolic rate drops significantly, reducing energy consumption and slowing down bodily functions. By inducing torpor in astronauts, scientists hope to minimize the psychological strain of isolation and the physical effects of microgravity.

In this state, astronauts could sleep through most of the journey, waking periodically for health checks or maintenance tasks. This not only reduces the mental challenges of long space missions but also cuts down on the resources needed, such as food, water, and oxygen. In many ways, this application of sleep is a form of psychological time travel, allowing astronauts to perceive the journey as taking far less time than if they were awake and conscious the entire way.

Suspended Animation and Torpor: The Science Behind It
Suspended animation, or torpor, is a state that resembles hibernation in animals. During torpor, the body's metabolic processes slow down, lowering the amount of energy necessary to sustain life. NASA and other space agencies are researching how to safely induce and maintain torpor in humans for extended periods, with the goal of making long-duration space travel more feasible.

In a state of suspended animation, astronauts would experience a journey to Mars or beyond as though they were taking a long nap. Upon waking, they would find themselves at their destination, having bypassed the months of monotony that would otherwise accompany interplanetary travel. This concept of sleep as a tool for space exploration brings humanity one step closer to the dream of time travel, where the boundaries of time and space are softened, and the limits of human exploration are expanded.

Prospects: Sleep as a Tool for Space Exploration
While the technology to induce torpor in humans is still in development, the concept holds great promise for the future of space exploration. By using sleep-like states to reduce the psychological and physical challenges of long space missions, we may one day make interplanetary travel more practical and less taxing on the human body and mind.

In the future, astronauts may embark on journeys to distant planets by entering a sleep pod, drifting off to sleep, and waking up at their destination. This vision of space travel, where sleep serves as a vessel for psychological time travel, brings us closer to realizing the dream of exploring the cosmos. Whether through dreams that allow us to mentally travel through time or through sleep-induced states that help us endure the long journeys of space exploration, sleep may hold the key to humanity's future in the stars.

CHAPTER 16: CONCLUSION

Recap of Key Concepts

In this subchapter, we will revisit the fundamental concepts introduced throughout "Time Unraveled: A Simple Guide to Quantum Travel." Understanding these key ideas is essential for grasping the intricacies of time travel as it relates to quantum mechanics and Einstein's theories of relativity. By revisiting these concepts, readers can better appreciate how the threads of science weave together to form a cohesive understanding of time and space.

One of the most crucial ideas to grasp is the theory of time as a dimension rather than a linear progression. Traditional views hold that time moves forward in a straight line, but in the framework of relativity, time is intertwined with the three spatial dimensions, creating a four-dimensional space-time fabric. This implies that time can be affected by factors such as speed and gravity. The quicker an object travels through space, the slower it moves through time, a phenomenon known as time dilation. This essential principle lays the foundation for recognizing how travel through time might be possible.

Another key concept is the role of quantum mechanics in the discussion of time travel. Quantum mechanics introduces the idea of superposition, where particles can live in multiple states at once until they are observed or measured. This principle challenges our classical understanding and suggests that time may not be as fixed as we perceive it. The possibilities of parallel universes and alternate timelines arise from this quantum perspective, offering intriguing scenarios for how time travel could function. These ideas push the boundaries of our imagination and scientific inquiry, suggesting that time travel might not only be theoretical but also a potential reality under certain conditions.

Einstein's theories of relativity further enrich our understanding of time travel. His general theory of relativity posits that massive objects can warp space-time, creating gravitational fields that influence the passage of time. This concept is essential for any practical application of time travel, as it implies that one could theoretically navigate through time by manipulating these gravitational fields. Additionally, Einstein's famous equation, $E=mc^2$, highlights the interchangeability of energy and mass, laying a foundation for understanding how energy could be harnessed to enable time travel.

Lastly, the implications of time travel extend beyond the realms of physics and into philosophical considerations. The idea of altering past events or encountering different timelines raises questions about causality, free will, and the nature of actuality itself. As we contemplate the potential for time travel, it becomes essential to consider not just the scientific mechanics behind it, but also the ethical and philosophical ramifications. By synthesizing these key concepts, readers can appreciate the complexity and wonder of time travel, paving the way for further exploration into the possibilities that lie within the fabric of our universe.

The Ongoing Quest for Understanding Time

The quest to understand time has captivated human minds for centuries, intertwining philosophy, science, and even art. From ancient civilizations who tracked celestial movements to modern physicists grappling with the complexities of quantum mechanics, time remains one of the most profound mysteries of the universe. As we delve deeper into the nature of time, we realize that it is not merely a linear progression from past to future, but a dynamic entity that may behave differently under various conditions, especially when examined through the lens of relativity and quantum theory.

Einstein's theory of relativity reformed our understanding of time by introducing the concept that time is not a constant. According to relativity, time can stretch and compress depending on the viewer's speed and gravitational field. This means that a person traveling at near-light speed would feel time more slowly than someone at rest on Earth. Such implications challenge our everyday perceptions of time as a uniform experience shared by all. This phenomenon, often referred to as time dilation, raises intriguing questions about how time functions in the universe and whether it can be manipulated.

In addition to relativity, quantum mechanics supplements another layer of intricacy to our understanding of time. At the quantum level, particles exist in states that defy classical interpretations, and events can appear to occur out of sequence, leading to a sense of timelessness. Concepts like superposition and entanglement suggest that particles can be interconnected across vast distances, blurring the lines of cause and effect. These quantum peculiarities invite speculation about the very fabric of time and whether it can be traversed or altered, igniting the imagination of scientists and dreamers alike.

The pursuit of a unified theory of time that combines the principles of relativity and quantum mechanics persists as one of the greatest argues in modern physics. Researchers are exploring various models, including string theory and loop quantum gravity, in hopes of reconciling these two realms. Each theory proposes different mechanisms for understanding how time operates at both cosmic and subatomic scales. The ongoing research not only seeks to answer fundamental questions about time but also opens the door to potential breakthroughs in technology and our understanding of the universe.

As we continue to explore the nature of time, it is essential to engage both the scientific society and the public in this fascinating discussion. Understanding time is not solely an academic endeavor; it touches on our deepest curiosities about existence, the universe, and our place within it. The ongoing quest for understanding time invites us to ponder profound concepts and encourages us to envision a future where the mysteries of time travel and quantum manipulation might one day become a reality.

Final Thoughts on Time Travel and Quantum Mechanics

As we close our exploration of time-travel and quantum mechanics, it is important to pause and reflect on what these ideas tell us about the possibility that the Universe is different from how we otherwise perceive it. Time-travel is one of the oldest and most creative tropes of sci-fi. But at its heart are the contortions of quantum mechanics – and the ideas of brilliant scientists that could soon displace those contortions entirely.

It changes our understanding of the basic structure of nature, introducing the possibility that a particle can have two or more definite states – while maintaining each state's distinctness – with a degree of intensity determined by the probability of finding it in the corresponding position. This principle, known as superposition, tells reality is more complex than simply from our common, the concept of a quantum can imagine the past, the present and the future as three separate threads woven together as a state in this quantum tapestry of possibilities, with different probabilities attached to, what's probabilistic future becomes a causal past — and things get messy What might be the implications of this? For one, that the past, the present and the future could be entangled in the form of a quantum superposition. And it might also imply a violation of the notion of some distinct direction of time.

Einstein's theory of relativity is the substance for much of our modern thinking about time travel. Its equations show that time is not an unalterable and immaterial dimension, but a dynamic one that can stretch like elastic tied loosely around the universe. The connection applies to time dilation, in which observers at different gravitational fields, or moving at different speeds, experience the flow of time differently. Objects closer to a large gravitational source will appear to pass through time more slowly. Or maybe time travel is not possible because the very flow of time is linked to the structure of space itself, rather than some mysterious entity.

While the theory could eventually allow practical time travel, none of this is anywhere near current technology or our current understanding of physical laws, so gratifying as it might be to time-travelling physicists, and the odd businessperson, its use is still limited to our theoretical toolbox. Today's quantum results lead to enough innovative ideas and work that we probably will gain a better understanding of what is possible and impossible in time travel over the coming decade – but, for now, at least, we will continue to walk rather than ride into our future.

Thinking about any of this – light cones, many-worlds dynamics, the measurement problem, Schrödinger's cats – brings us face to face with perplexing scientific concepts, riddled with philosophical significance, that are likely to provoke just as much questions about existence, the nature of reality and the limits of human free will as anything we might come up with as an answer. We need not fear that in pursuing such lines of thought we are raising questions that our science cannot one day provide us with explanations of. Quite the opposite. If we are willing to go there, any such 'break through to the other side' – from relativistic 'spacetime' physics to our subjective experience of 'time' – is bound to force us to confront, every bit as much as it always has, an 'other side' that is very much of us. It is, after all, a continuation of an age-old curiosity about the nature of the Universe and our place within it.

www.ingramcontent.com/pod-product-compliance
Lightning Source LLC
Chambersburg PA
CBHW070410230526
45471CB00006B/2730